NIAMPA A. S. Boukari

Methods and results in the management of natural resources

NIAMPA A. S. Boukari

Methods and results in the management of natural resources

of the Kou watershed in Burkina Faso

ScienciaScripts

Imprint

Any brand names and product names mentioned in this book are subject to trademark, brand or patent protection and are trademarks or registered trademarks of their respective holders. The use of brand names, product names, common names, trade names, product descriptions etc. even without a particular marking in this work is in no way to be construed to mean that such names may be regarded as unrestricted in respect of trademark and brand protection legislation and could thus be used by anyone.

Cover image: www.ingimage.com

This book is a translation from the original published under ISBN 978-620-3-44221-2.

Publisher:
Sciencia Scripts
is a trademark of
Dodo Books Indian Ocean Ltd. and OmniScriptum S.R.L publishing group

120 High Road, East Finchley, London, N2 9ED, United Kingdom
Str. Armeneasca 28/1, office 1, Chisinau MD-2012, Republic of Moldova, Europe
Printed at: see last page
ISBN: 978-620-5-72640-2

NIAMPA Boukari

Socio-economist of development

+226 76 59 57 80/62 94 79 35

niampaboukary@yahoo.fr

Free image

The problem of sustainable management of natural resources and associated ecosystems in the Kou watershed

Which methods for which results?

Comparative analysis of natural resource management methods: from customary practices to modern colonial and post-colonial practices.

(Excerpt from a research paper on methods of managing natural resources and associated ecosystems)

Introduction

The methods and practices conveyed are at the heart of natural resource management. They constitute the main object of this research. They will therefore be identified and studied in order to identify and compare their strengths and weaknesses with the current and future challenges related to the sustainable management of natural resources.

The research was conducted in the Kou watershed. It looked at both traditional and modern methods of natural resource management in the watershed.

1.   Traditional methods

> *This part of my research is based on the writings of Professor Doti Bruno Sanou, researcher and founder of the African Center for the Cultural Practice of Development (CAD). In my reading I came across these writings that I found very eloquent with what I was trying to show. Here, the Sogo practice of forest and space management has been written in marvels.*

Also, methods such as those inspired by the Bobo Mandarè custom have been used in the management of natural resources in the Kou watershed.

1.1.   The Sogo concept of natural resource management

The word "bush" can be used in the Madare culture to describe the concept of "Sokogo" or "Sakuè" in the Vore dialect, which is to be implied in the concept of "Sogo" or "Saga" used to designate nature in its wild state before human intervention.[1]

1.1.1. Perception and principles of forest management

---

[1] SANON, G., *quoted by D.B. SANOU, in Coutumes et aménagement des forêts au Burkina Faso, Contribution de la coutume bobo-madarè à l'aménagement des forêts de Dindéresso et du Kou dans la commune de Bobo-Dioulasso p.38*

The traditional perception of the forest made it at the same time a space of potential insecurity, because it could harbor dangers of various origins: attacks of the enemy or predators, but above all of human security, refuge of protective spirits and sources of socio-cultural and economic activities to be preserved and shared with the generations to come. Generally sheltering a watercourse, the forest is also a symbol of purity, a holy place that must be spared from the defilements of human origin.

Thus, according to Doti Bruno SANOU (2004)[2] , the management of the forest aims to make it :

- harmonious to serve as a place of security, fertile and fertile for various human activities such as gathering, hunting, agriculture, travel ;

- to purify the forest from desecration or pollution of anthropic origin;

- to serve as an ecological conservation area as indicated by the Mandare adage: "Every river needs its forest".

- The forest is thus seen as a natural support for the infiltration of water, even the formation of a watercourse;

- the forest is also a breeding ground and a nursery since the fauna and the flora are protected there (a place of refuge and reproduction of the animal species but also of dissemination, germination, expansion of the species).

For it to play this role of social regulator and sources of life, the forest must benefit from a particular care by the members of the society. They must adopt behaviors that contribute to protect it and to preserve its biophysical and functional integrity.

Thus, a regulation based on Sogo values and principles, found in customary law

---

[2] SANOU, D.B., (under the direction of) *Communautés villageoises delaprovince du Houet et projet BKF/007. PAFDK. Un exemple de partenariat pour l'aménagement des forêts classées de Dinderesso et du Kou,* CAD, 2004 p.29, 30, 31

("landa"), whose application is done in a collegial manner by the "Sogo vo", the "Kirevo", the "Dovo" and the "Yèlèvo", has provided for precise laws regulating the activities of villagers in the "bush" concerning wood cutting, gathering, hunting and fishing.

This regulation makes these practices factors of preservation, protection, and conservation of natural resources. Thus, bushfires are only allowed after the "fogosiokiè" or sacrifice of the first fruits of the new harvest, and they are forbidden after the "Sanibige" or end-of-harvest celebration and return to the village. The primary purpose of the bush fires is to create a safety belt around the huts in an environment where the grass is very high. A safety belt against wild animals, reptiles or even criminals.

The same applies to hunting, gathering, fishing and access to rivers, which are strictly regulated and monitored.

Violation of these directives exposed the offenders to sanctions, including the obligation to pay for sacrifices of expiatory and propitiatory reparations intended to calm the spirits of the bush.

1.1.2. Principles of soil/land resource management

God created the heavens and the earth and from it he shaped man according to certain religious and African beliefs. The earth is therefore sacred, they say. It is a gift from God. The first occupants are by the will of God. They own it by delegation and must ensure its good governance. The land is therefore only accessible from an owner who ensures the security of its material existence. The latter can transfer it under certain conditions: lineage inheritance, solidarity with other families or for reasons of hospitality on behalf of friends and foreigners welcomed in the locality and who are sponsored by a host.

The technical and organizational management of the soil/land resource is governed by rules and principles that are known, binding and accepted by all. Access to and use of the land are subject to well-known and accepted conditions. The granting of land is often symbolized by a sacrifice (rooster, chicken, cola, etc.). Once the land is acquired, the beneficiary is required to participate in the ceremonial rituals or traditional festivals dedicated to it and organized annually. This may involve gifts in kind (grain, chickens, dolo, etc.).

At the beginning or at the end of the year, each farm manager is required to make sacrifices in the form of offerings in order to renew the various alliances linked to the enjoyment of the usufruct.

The allocation of a field to a person requires, in the Madare mentality, a knowledge of how to live in a community based on respect for others, on solidarity in times of trial, on mutual aid in work and on an ethic that must be rigorously observed (SANON, G, 128)[3] .

The exploitation of the land is accompanied by compensatory practices to ensure its biophysical integrity for a long time. Thus, fallow practices allow fields to breathe for 3 to 15 years and allow the soil to recover. At certain times, it is necessary to stabilize the soil against sheet erosion, notably with fonio, whose stem is an excellent construction material (banco reinforcement). It is used in adobe construction, one of the secrets of the durability of traditional buildings that can last 100 to 200 years.

Bush fires are either prohibited or reasonably organized to allow for land renewal: burning in newly cleared fields, early fires to protect soils from bush fires.

The treatment of gullies is mainly done at the scale of the agricultural plot by vegetation from andropogon, tonolé, or by mechanical means such as branches.

The cultivation method of staking and mounding used for this purpose, favors

---

[3] SANON, G., *Le monde comme dehors et dedans, Essai sur la philosophie madarε*, Thèse de3ᵉ cycle, Strasbourg, 1980. Op.cit, P. 128

drainage, but also the conservation of moisture and organic fertilization.

Technical itineraries, especially the association of low crops with high grasses, are also techniques that help reduce soil compaction as a means of conserving the soil.

The process of fertilization by leguminous plants, early bean in mixed cultivation proved to be successful in this matter in the past.

1.1.3. Water resources management

In terms of water resources management, the "Sogo" has regulated gallery forests, riverbanks and springs as areas of absolute classification, with a total ban on all human activity.[4]

This practice corresponds to the spirit that rivers and forests are the same cosmic entity because where there is a dense forest, there is an integrated river.

This is why, according to the Bobo Mandare concept, "each forest requires its own watercourse" (SANOU, D.B; p.67)[5] .

The management of water resources is also made up of prohibitions, as is the case with the forest. Violation here can expose oneself to the cry of distress of the elders or to the aggression of a crocodile or a hippopotamus, representing the same elders.

In case of violation, opportunities for atonement and propitiatory reparations are arranged to calm the river spirits.

In addition, certain acts, such as crossing the marigot in shoes or cutting trees on the banks, are prohibited. Access to water sources, although common law, is effectively regulated.

---

[4] SANOU, D.B., *Un exemple de partenariat pour l'aménagement des forêts classées de Dinderesso et du Kou,* CAD, 2004 p.29, 30, 31. Idem, p.67
[5] SANOU, D.B., Idem, p.32

Thus, any practice, which can distort the property of water is formally prohibited so that it was possible in the case of the forest of Kou to see any object at the bottom of the water.

Any violation of other prohibitions, such as sexual practices in the bush, is strictly prohibited. If a couple lives in the bush (temporary farming hamlets), they are obliged to return to the village. Homicide or the spilling of human blood, or t h e delivery and burial of placentas, were prohibited[6] .

1.1.4. The organization of the space

The planning of the village territory "Kibèko" or "Sagako"[7] was conceived according to the availability and the utilitarian, strategic, socio-cultural, economic and ecological character of natural resources. Thus, in addition to being economic objects, natural resources were personified and the relationships between humans were such that they formed a harmony with their environment[8]

.

The location of each village is characterized by a main river (the
"This is due to the existence of the bush altar (the "Saga") and the mountain area (the "Tolo"). The bush altar is dedicated to crop fields (agricultural production, grazing)[9] . The presence of the mountain expresses the manifestation of the force of nature, that of the river is a sign of the dynamism of the world, the vitality of nature.

The development of the village territory for agricultural purposes is done according to lineage affinities. This is divided into zones of cultivated fields,

---

[6] This implies that women were to be spared from going to the bush.
[7] SANOU, D. B., (Under the direction), *Coutumes et aménagement des forêts au Burkina Faso, Contribution de la coutume bobo-madarè à l'aménagement des forêts de Dindéresso et du kou dans la commune de Bobo-Dioulasso*, p.38

[8] SANOU, D.B., Idem, p.33
[9] Crops, water supply points for drinking and animals, etc.

"lo" or

"laga", belonging to the different large families constituting the village according to the paternal lineage "Konkuma", headed by the oldest "konsonmlalo". These fields are called "konnoma laga" or field of the sons of the same household. This field is a common property of the family and cannot be sold. It can be exploited either in "Foroba" (collectively) or in "Zakanè" (individually) or both possibilities at the same time[10] .

In addition, the territory is organized in order to associate it with practices that respect its physical integrity[11] . On each terroir (Kôlo) is applied with land charters. The production lands are managed with codes of exploitation according to customs and traditions. Each Kôlo has its own specific prohibitions (access to certain products), in the form of a charter that organizes daily activities. The community of operators of the said space is created independently of the village origins. This community ensures that everyone behaves in accordance with the charter and the general prescriptions of the Sogo. This is likely to maintain a true spirit of community solidarity around natural resources.

Other types of development participate in the organization and management of the village space. These are mainly :

-       the village bocage[12] (Kankiré or Kagniré): this is an annular reservation around the built-up area of the village, intended to receive garbage, including defecation, the grazing of small ruminants, fields of minor crops such as tobacco, etc. It is also a reservation for building space. It is also a reservation for building space.

---

[10] SANOU, D. B., (Under the direction), *Coutumes et aménagement des forêts au Burkina Faso, Contribution de la coutume bobo-madarè à l'aménagement des forêts de Dindéresso et du kou dans la commune de Bobo-Dioulasso*, p.39

[11] SANON, K.B., KONATE, Exposé sur les Implication des chefs de terres et des autorités religieuses dans la conception et l'adoption d'un plan d'utilisation des terres, 2008

[12] Area where fields are enclosed by hedges or rows of trees and where the habitat is arranged in farms and hamlets.

-       the source of drinking water supply: generally the village is located on the edge of a watercourse with perennial sources for drinking water supply.

They are the subject of specific code including aquatic fauna.

-       the living space (Kourou-cité): the Bobo city is a compact block of buildings with socialized spaces (areas for events, altars for worship, etc.). The city is a sub-space of the "Sogo" with particular rules. It is materialized by an altar (a conical stele with an oval base visible in every Bobo village, celebrated annually).

The creation and protection of sacred spaces[13] and places of worship also participate in this spirit of development: the sacred tree, symbolizing in a general way the soul of the village *(Sogo, sounsouyiri), is* generally forbidden to be cut. It is even respected to the point of often receiving a funeral. Sacred woods play an important role in the social and cultural organization of each

 locality. This cosmic arrangement of natural space translates a whole vision of the world, the concept of life of the Bobo Madarè[14] .

Clearly, traditional methods and practices are based on rigorous planning of interventions for the exploitation and protection of natural resources and are surrounded by known and followed rules that minimize uncontrolled aggression on these resources. All these things contribute to their sustainability. And as the Bishop Emeritus of Bobo-Dioulasso, Monsignor Anselme Titianma SANON, points out, "*the indigenous population had a basic management system that took into account measures or interests that imposed respect for the rhythm of nature[15] .*

---

[13] The sacred space can be on the mountain, by the river, under the big trees of the village or in the bush

[14] SANON, G., *Le monde comme dehors et dedans, Essai sur la philosophie madarɛ*, Thèse de3ᵉ cycle, Strasbourg, 1980. Op.cit, P. 128

[15] SANON A. Titianma, Bishop Emeritus of Bobo-Dioulasso, Interview conducted on November 06, 2012, from

These traditional methods contain principles and values that correspond to the sustainability requirements of the resources in question. These principles and values should have been taken into account by the colonial and post-colonial methods, which need a socio-cultural foundation in order to be better appropriated and applied by the different communities that place their trust in them.

2.   Modern methods

They have evolved from colonial methods to post-colonial methods characterized in the most extreme cases by repression, punishment and prohibition, a kind of exclusion of the population from the management of the natural resources of their living space, inherited from the ancestors.

2.1.   olonial methods of natural resource management

The colonial methods proceeded mainly through the creation, development, conservation and management of natural forests that the colonial administration exploited for industrial, economic, commercial and tourist purposes. This type of management, if it aims at economic purposes, also participates in the protection of the forests against the aggressions of mainly anthropic origin. Regulatory measures are applied and these are imposed on local communities. The latter were rarely consulted in the management of these forests and were expropriated from their natural heritage. Fundamentally (CCFM, 1992)[16] , "sustainable forest management" aims "to maintain and enhance the long-term health of forest ecosystems for the benefit of all living things... while providing good environmental, economic, social and cultural opportunities for present and future generations. Thus, the natural forests of Dindéresso and Kou were

---

4 to 5 p.m. at the Sénoufo Cultural Center

[16] Canadian Council of Forest Ministers (CCFM), 1992.

classified[17] and put under colonial management for at least forty years before the proclamation of the country's independence. Development actions were undertaken such as the development of the guinguette as a place of rest and recreation previously intended for settlers. The two forests were classified in accordance with the orientations of the colonial forestry policy[18] stipulated in the decree of August 3, 1928, organizing the Water and Forestry services in Upper Volta.

The Dindérésso Forest was classified by decree 20/n°99 of 27 February 1936. It covered an area of 7000 ha before being enlarged by decree n°3006/SE of August 26, 1941 by 1500 ha, bringing its area today to 8500 ha. The first vocation of the classified forest of Dindérésso is the forest management for the production of firewood.

As for the Kou forest, it was classified by decree n°227 of 22/1/51 of January 30, 1951. Its main vocation is the conservation of biological diversity and the protection of the springs of the guinguette. A real ecotourism site, it shelters sacred woods venerated by the riparian villages.

The colonial management of the forest areas that have been classified has helped to anticipate the phenomena of degradation due to human action. As a result, these two forests were able to survive and now serve as an example of imposed natural resource governance[19] . This practice is all the more salutary since classified forests today play an important role in the provision of ecosystem services to rural and urban populations. Thanks to the existence of this forest, the natural springs in the Kou classified forest have been safeguarded. Even if, due to bad

---

[17] The two forests were classified in accordance with the orientations of the colonial forestry policy stipulated in the decree of August 3, 1928, organizing the Water and Forestry services in Upper Volta

[18] The forestry policy of the colonial administration aims at the conservation of natural woods, the limitation of deforestation, the creation of vegetal climatic barriers

[19] Throughout the country, there are 78 classified areas, 74 of which were created between the 1930s and 1950s (colonial period) and only 4 after (independence period). Between the 1930s and 1950s, there were 74 classified forests for about 20 years, i.e., an average of 4 CF per year, whereas from 1950 to 2011 (54), there were only 4 classified forests, i.e., an average of 0.08 CF per year.

human practices, their number and capacity to emerge are decreasing, these natural springs constitute one of the sources of drinking water supply for the city of Bobo-Dioulasso by the National Office of Water and Sanitation (ONEA).

These forests also act as a safeguard against the advance of desertification in the area. Their management provides important benefits for the fight against poverty and local economic and social development.

Despite the relevance and effectiveness of the colonial method in guaranteeing the sustainability of the natural resources of this basin, the application of this method has not always been to everyone's taste. It has often been accused of being a source of usurpation of rights and exclusion of those entitled to the developed areas.

Table N°5: Situation of natural resources conservation areas in Burkina Faso

Vocation	Number	Area (ha)	of ter Nat
Forest classified as a wood reserve	53	632 783	15,56
Classified forest with wildlife vocation	12	412 250	10,14
Partial Wildlife Reserve	5	413 247	10,16
Total Wildlife Reserve	3	306 482	7,54
National Park	2	458 380	11,27
Silvopastoral and partial wildlife reserve	1	176 4367	43,39
Biosphere Reserve	1	16 970	0,42
Game Ranch	1	62 162	1,53
TOTAL	**78**	**4 066 641**	**100,00**

- *Source: MEE (1996)*

For others, the colonial method ignored or even alienated the positive customary practices that were in place in the natural forests and that were well respected by the community members. This exclusion of traditional values, principles and institutions was counterproductive from the point of view of ownership and participation of the population in the management of post-colonial forests.

2.2. Modern postcolonial methods

They are mostly inspired by colonial methods of natural resource management. However, they have changed in response to the socio-political, economic and environmental developments that have characterized the context of environmental and natural resource management in the world in recent years.

In the Kou watershed, the investigations identified and analyzed several methods deployed by the various stakeholders.

a) *Administrative and institutional methods*

They are mainly used by the general administration and technical services. They give priority to consultation and dialogue structured through the regulatory frameworks provided for in the various texts[20] or the thematic frameworks set up to deal with specific sectoral issues. Consultation undoubtedly makes it possible to involve stakeholders in the search for and application of solutions that ensure the sustainable and efficient management of the BVK's natural resources, particularly through the synergy of action of the actors. The best known frameworks are the regional consultation frameworks (CCR), provincial consultation frameworks, regional and provincial land use planning commissions (CRAT), communal consultation frameworks, and other thematic or multi-sectoral frameworks dedicated to addressing specific sectoral issues. Each framework has a number of sessions to hold during the year. In principle, this number is 4 for institutional consultation frameworks such as the CCR, CCP and CCCo. For the thematic frameworks, the number of sessions depends on the issues to be dealt with, for example, the sessions of the CRAT, the CPAT or the CLE or others. The management of natural resources in the watershed has, on many occasions, led to the holding of thematic sessions. The last one to date is

---

[20] The HWRA, the TMC, the NRHP, the SCADD, etc.

that of the CLE-Kou devoted to the development of an emergency action plan and a short and medium term action plan. The debates within these frameworks on the management of natural resources in the Kou watershed were often fascinating because these resources, in addition to being sources of divergence in methods, are the object of several sectoral covetousness. The sessions have even often mobilized the highest authorities of the State. These exchanges and discussions have always resulted in proposals for action, recommendations and decisions to be taken. However, it must be recognized that in most cases, the proposed actions, decisions or recommendations made have not been followed up. This void in the application of solutions to the degradation of the basin's resources is not likely to contribute to their sustainability.

b)   *The methods of the deconcentrated technical services of the state*

In the context of the Kou watershed, they are almost the prerogative of the technical services of rural development and result in the implementation of a system of supervision of the actors at the base. On the agricultural level, they consist mainly of the supervision of agricultural producers in the adoption of agricultural and technological practices that increase agricultural production at the plot level. The trend today in this area is to use improved seeds or short-cycle varieties to overcome the lack of rainfall and fertile land in the watershed.

Although indispensable to producers lacking technical agricultural instruction, the methods deployed by the agricultural services have remained productivist. Issues related to the sustainability of natural resources have not often been associated with development and agricultural production actions. In other words, the equalization between the productive capacities of natural resources and agricultural production needs has never been fully addressed.

On the pastoral level, the supervision methods consist of sensitization, training, support and advice to the producers of the sector in their breeding activities in all its diversity: breeding of small and large ruminants, economic breeding, traditional breeding, professional breeding. In this sector, some technological innovations have been introduced such as the breeding of crossbred hens, the improvement of breeds, artificial insemination, and techniques for the collection and conservation of natural fodder. In the Kou watershed, a group of breeders are engaged in this last practice thanks to the support of PAFDK and PAGREN. Although they have evolved over time, the methods used in the sector have not allowed the herders to participate in the sustainability of the resources they exploit for their livestock and, above all, to operate an agriculture-livestock-environment association that optimizes the economic and ecological functions of these natural resources. In addition, the methods of development and management of pastoral areas, although concerned about the scarcity of food sources or the difficulties of accessing them, have not been able to resolve at the watershed scale the issues related to the stabling of animals, the modernization of livestock by orienting it towards the practices of an economic livestock farming of competition, competition and creation of planned wealth. The controlled grazing in experimentation in the classified forest of Dindérésso is one of the examples that participate in the planning of grazing, but the problem remains because its adoption and application by the various actors still has a long way to go.

At the forestry level, the technical method is currently limited to surveillance, repression and verbalization activities both inside and outside classified forests. This surveillance, beyond the fact that it is insufficient and not permanent in time and space, rarely combines verbalization methods with awareness-raising, communication and training to influence bad behavior and reduce acts that are harmful to the integrity of natural resources in the Kou watershed. To this must be added the growing complicity between certain foresters (*according to several testimonies and even observations*) in charge of the protection of the forest and not of the protection of their selfish interests from the forest, and certain individuals lacking alternatives and little aware of the consequences of the acts that they compromise the existence of these natural resources and the livelihoods of thousands of people living from their exploitation.

The scheme below is a desired approach for the committed forester. Apart from that, it is more a source of insecurity than of security for the forest in Burkina Faso.

Figure 1: The committed forester's approach

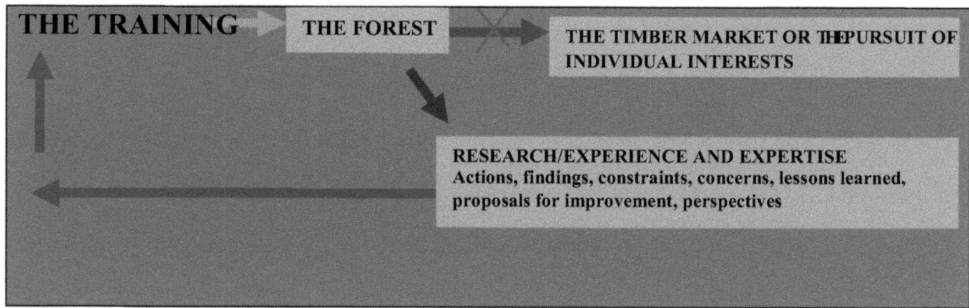

As a result, the methods used by the technical services of the State as a whole, beyond the constraints caused by mobility and the weakness of means, remain classic compared to the stakes and emerging challenges in the management of natural resources such as climate change and the sustainability to be sought.

Apart from these practices that are not very concerned about the future of these natural resources and with them that of several generations, the lack of synergy of action and sometimes even the contradictions between development sectors in relation to the attitudes to be given to users indicate the need to work on changes in paradigms and conception of the natural resource of the basin which must be used on a sustainable basis for economic and social development that respects its physical and functional integrity. For example, the interpretation of the regulatory 100 meters, although not subject to any ambiguity, remains a stumbling block between the agricultural services, which favor agricultural production in spite of all, and the environmental services, which defend the exploitation beyond the regulatory 100 meters. And the breeding in that, how to allow the animals to reach the sources of water to drink there. There is necessarily a need to work together within the framework of the same vision.

In a perspective of sustainable management of natural resources that implies taking into account the divergent interests of users, the only issue that is worthwhile is to find a consensual development and management scenario that guarantees the physical and functional sustainability of these natural resources. In this way, alternative solutions to illegal activities can be imagined, because in any case, only force can ensure the regulations in force.

As a result, the methods used by the technical services to manage the natural resources of the watershed, even when brought into contact with socio-political, economic and environmental requirements, have not sufficiently internalized the practices that respect the dimensions of sustainable development: ecological, social, economic and cultural.

These are shortcomings that stem from a lack of governance that brings everyone together and commits them to the same goal.

c)   *Methods of approaching projects and programs*

Natural resource management through physical development is a system for controlling resource use so as to avoid waste and to optimize resource use for maximum benefit (Collins, 1992)[21] .

---

[21] Collins, P. H., 1992. *Dictionary of Ecology and the Environment,* Universal Book Stall, New Delhi, India, 117 pp. Quoted by OUEDRAOGO, B., in aménagement des forêts et lutte contre la pauvreté au Burkina Faso

➢ **Forest management methods in Burkina Faso**

The management of forest resources in West Africa is sequenced into three major periods, pre-colonial management, management during colonization and post-colonial management.

Pre-colonial management was governed by traditional principles, rules and values that have proven effective[22] over time. These methods are increasingly attracting interest, justified by the repeated failures of natural resource management by so-called modern methods.

The colonial management system aimed mainly at delimiting and marking the boundaries of natural forests for protection purposes, but also for exploitation to satisfy the economic, commercial and industrial interests of the colonial administration. It dates back to 1933 when several classified forests were created, including the Kou and Dindérésso forests in the Kou watershed.

The period of the 1960s (the "independence" period), characterized by forestry policies inherited from colonization, where conservation policies were enacted by the State with little participation by the populations. This method thus took on a centralized and technical character (Ribot, 2001)[23] . The argument based on dialogue, negotiation and consensus used by traditional methods has given way to the argument of force and repression. The failure of most natural resource management policies such as forests is due in part to this centralization and technicism and the exclusion of community opinion and voluntary participation.

The failure of post-colonial forest resource management methods in terms of physical, social (population participation) and strategic (appropriation and

---

[22] And as the Bishop Emeritus of Bobo-Dioulasso, Monsignor Anselme Titianma SANON, points out, *"the indigenous population had a basic management that took into account measures or interests that imposed respect for the rhythm of nature*

[23] Ribot J.C., 1998. "Theorizing access: Forest profits along Senegal's charcoal commodity chain." *Development and Change*, 29, pp. 307-341. Cited by OUEDRAOGO, B., in Gestion des forêts et lutte contre la pauvreté au Burkina Faso

sustainability of results and achievements by local populations) results in the political authorities and technicians reviewing their approach, in particular by favouring an approach that effectively involves and empowers the populations. This trend of opening up to local actors was strongly recommended by the United Nations Conference on the Environment and Natural Resources held in Rio in 1992. The latter had also detected the inefficiency of the system due to its technical and centralized nature, maintained by the technical services despite socio-political, economic and environmental changes such as climate change and the great droughts of the 1970s.

The principle of participatory forest management was established in 1986. It has been translated in the field by the implementation of several participatory approaches to natural resource management[24] throughout Burkina Faso. The management of natural resources in the Kou watershed has also been the subject of participatory planning and management, particularly around and in the classified forests of Dindérésso and Kou.

The participatory approach favored in the management of these two forests aims to improve the ownership of the approach by the actors concerned, mainly the forest management groups and the forest administration, and thus facilitate the continuation of management activities and the sustainability of the gains already made.

However, this participatory approach, which is sometimes well appreciated because of the results visibly achieved, suffers from a low level of ownership by the various direct beneficiaries (populations, GGFs) and indirect beneficiaries (technical services, local authorities).

In fact, the majority of the local populations, who legitimize their ownership of

---

[24] The "8000 villages, 8000 forests" approach, which evolved into "one department, one forest"; "one school, one grove", which evolved into a "one school, one orchard" program; participatory approaches, including project approaches, integrated development approaches, and "land management" approaches

these forests, are not very involved in this management, despite the regulatory measures that now recommend it. The representation of the populations by the Forest Management Groups (FMGs) in projects and programs has often been misunderstood. The GGFs have been repeatedly accused of being friends of projects and programs that grant them special favors for tasks that are not very visible at the village level. This perception of the role of the GGFs naturally undermines the level of participation of the populations in forest management, if not sabotaging it.

> **Hydro-agricultural development methods**

Hydro-agricultural developments aim to facilitate the mobilization and planned use of water to meet the ever-increasing production needs. There are several types, such as irrigated perimeters, micro-dams, market gardening and pastoral wells, boreholes, etc. Development actions are most often preceded by diagnostic, socio-economic and economic studies and supported by a mapping of available resources.

The implementation and valorization of these achievements is often based on established management rules that engage and make resource users responsible for their respect. In the Kou watershed, the VREO and GEeau projects have used these methods.

Even though it produces tangible physical results, this method remains technically advanced in relation to the users' level of conception. Many of them are unable to read and use a map as a first support for planning and monitoring-evaluation of the actions promoted.

Thus, technical accessibility and social appropriation remain the challenges to be met by this method if it is to bring about a change in the way users of water resources see and act. Apart from that, the question of the sustainability of the hydro-agricultural works carried out is a major issue if anticipatory measures are not taken to ensure continuity in their management, guarantee their social appropriation and optimize their use in a sustainable manner.

> ### ➤ Conservative water and soil management methods

It is about developing and adopting land and water use practices that help maintain their biophysical and functional integrity. It is a method that aims to minimize practices that degrade soils and disrupt waterways and to promote techniques that conserve and enhance ecosystems. This method (Roose 1984)[25] defines a system of exploitation that allows for the conservative management of available water and soil fertility, mainly by allowing for the selection of efficient devices that spread out the water, slow down the speed of its flow, slow down and disperse its energy, reduce its transport capacity, in short, allow for less destruction of the soil structure

In the Kou watershed, this method has been used practically by the BKF/012.PAGREN Project[26] during the last ten years. This method was adopted on the basis of the results o f a study that made it possible to diagnose and determine the root causes of water and soil degradation in the basin. The analysis of the causes has made it possible to identify possible solutions in terms of appropriate techniques and technologies. These solutions were shared and exchanged during a session of the Kou Local Water Committee (CLE-Kou). Thus, amendments were made as well as proposals to serve the development of a development and conservation management scheme for water and soil applied to the Farakoba sub-watershed. One of the proposals consisted in creating an Ad'

---

[25] Roose, E.J. 1984. Causes and factors of water erosion in tropical climates. Implications for anti-erosion methods. Tropical Agricultural Machinery 87: 4-18. Cited by Roose in Gestion conservatoire des eaux et de la fertilité des sols dans les paysages soudano-sahéliens de l'Afrique occidentale,

[26] Support project for the participatory management of natural resources in the Hauts-Bassins region, financed by the Luxembourg development cooperation

hoc Commission including the technical services of the State, the members of the civil society, the members of the territorial authorities and the communities represented by the various village development councils (CVD). The main mission of this Commission was to accompany the process of finalizing the proposed preliminary development and management plan and also to support the implementation of certain actions resulting from the solutions that were identified. The preliminary draft plan was transformed into a development and management plan by a special session of the CLE-Kou, which validated the various proposed actions and the techniques and technologies that would support their implementation. To begin with, and in order to make the different techniques and technologies accessible, effective and adapted, Field Schools were set up. This is a training-action, combining theoretical aspects with the use of techniques and technology demonstrations. Following these field schools, the participants, particularly the VDCs, having perceived the importance of water and soil conservation in the context of their agricultural activities, asked that it be institutionalized, i.e., carried out by permanent local organizations. The idea of creating a water and soil conservation management group in each village in the sub-basin. The latter is assigned missions of realization and animation of school fields, of mobilization of the actors on the good practices of GCES. Thus, 12 GESFA have been created in the 12 villages. Each group received technical assistance to set up a school field to popularize the practice or approach, which was later called *"reasoned practice in the field"*. This approach is highly educational and instructive. It enables the producer to observe the behavior of water and soil in his plot, to ask questions about natural and human factors and to find structured answers from the point of view of soil integrity conservation, water control and conservation. Thus, once the causes of degradation of his plot are established, the producer looks for the solutions that should be opposed to the phenomena. The final goal of this approach is to allow producers to find and apply appropriate solutions on the scale of their plot.

Thanks to this approach in the method, several hectares of degraded land have been recovered, and several others have been subject to anticipatory management. Thus, it can be said that the method has created a community awareness sanctioned by individual and collective initiatives in the development of crop fields in the spirit of water and soil conservation.

Although a given method or approach contains within itself the elements of its relevance and effectiveness, the fact remains that the results it has achieved have been insufficient to have a significant impact on human behavior and the physical restoration of the environment. The expectations of the populations have not been fulfilled to the extent of the promises made to them and to the extent of their commitment to undertake individual and collective actions that contribute to reversing the trend of degradation of the soil, land, waterways and ecosystems in the Farakoba sub-basin and in their own parcel. Some of the works carried out, such as the micro-dams, could be used for economic purposes and constitute an opportunity to better organize and make the village organizations more responsible for equitable, rational and sustainable exploitation.

The headlong rush of projects and programs and the absence of a spirit and a duty to follow up on the results and achievements that have been obtained thanks to the technical and financial support of lateral and bilateral partners is highly regrettable for a country such as Burkina Faso with very limited resources, where rigorous management of the means at hand must be the rule. Sometimes the reasons for the early termination of interventions or the systematic abandonment of the results they have achieved are to be found elsewhere [27] . Generally speaking, discussions on whether to stop or continue

---

[27] The reasons are often to be found in pettiness, jealousy, the pursuit of individual interests, abuse of power and bad faith in the choices and decisions that go with it. In this irresponsible mania, the populations are totally

interventions are held in the absence of the direct beneficiaries of the interventions.

The different methods studied, their strengths and weaknesses, and proposals for correction are recorded in the table below.

Table 6: Comparative analysis of the different methods

Reference values Sustainable management of natural resources (GDRN)	Forces	Weaknesses	Areas for improvement
Traditional methods			
Appropriate and enabling policy, legislative and regulatory framework	Existence of endogenous rules that are established, known, accepted and followed by the members of the community. Use of ethics and the spirit of intergenerational solidarity	Unwritten values that are poorly transmitted to new generations Weakly taken into account in modern NRM practices	Take into account the values and principles on which traditional NR management practices have been built
Planned and organized management of natural resources	Resource management is planned in time and space according to need. The concepts and practices of community-based land use planning and management are consistent with the values of shared and equitable governance of natural resources	Some prohibitions that limit canopy restoration initiatives in areas granted to third parties	Relaxing the provisions prohibiting actions on borrowed or leased land with rules of transparency, contractual and shared responsibilities in the management of the natural space
Emphasis on meeting economic and social needs on a sustainable basis[28]	The natural space is subdivided according to the economic vocations to be promoted production space (hut and bush fields) conservation space (bush altar), etc. Functional specification of spaces and the notion of conservation	Low level of production valorization	Introduce the concepts of investment and value added into traditional NRM practices

---

[28] Survival (satisfaction of basic food and sanitary needs, etc.), socio-economic (satisfaction of economic, cultural, educational and leisure needs, etc.), and financial (ability to save, invest, obtain economic and financial gains, wealth, etc.), and finally (attainment of a level of comfort, prestige, social image, etc.).

Reference values Sustainable management of natural resources (GDRN)	Forces	Weaknesses	Areas for improvement
Reference values GDRN	Forces	Weaknesses	Areas for improvement
Colonial methods			
Appropriate and enabling policy, legislative and regulatory framework	Definition of a rigorous legislative and regulatory framework that imposes social discipline in accessing and using	• Failure to take into account local economic and socio-cultural interests • Exclusion of traditional values • Technicism of the approaches used	Take into account the realities and social sensitivities and give new meaning to the role of local institutions in NR management
Planned and organized management of natural resources	• Creation, development and management of forests • Protection against degradation • Conservation of plant and animal species for economic, social and cultural purposes • Structural management	The exclusion of local institutions, rules and endogenous practices of natural resource management that give importance to concerted and consensual management and proceed by sustainable resource management is a source of weakness in this method.	Consider the values, principles, practices and spirit of planned natural resource management used by traditional methods. Involve local authorities and responsibilities in the management of the areas (forests) developed.
Emphasis on meeting economic and social needs on a sustainable basis	Development and management for economic, industrial and commercial purposes,	Exploitation intended to satisfy the economic, industrial and commercial needs of the colonial administration alone, to the detriment of the interests local communities	Take into account the economic interests of local populations in the form of compensation or the fruits of investments made in landscaped areas

Reference value in GDRN	Forces	Weaknesses	Areas for improvement
		Administrative and institutional methods	
Appropriate and enabling policy, legislative and regulatory framework	The spirit of consultation advocated by the various texts through the formally established consultation frameworks	Ineffectiveness of the frameworks marked by the low level of implementation of the decisions, recommendations or proposed solutions obtained from the consultations.	Work to make the consultation frameworks more effective from the point of view of implementing the fruits of reflection in the field, put in place a system for monitoring and evaluating the implementation of recommendations and decisions
Planned and organized management of natural resources	Existence sometimes of action plans resulting from the reflections	The action plans that have been drawn up have been implemented at a very low level, if at all, due to inadequate mobilization of the actors and resources needed for their implementation	Establish a strategy for mobilizing the human, material and financial resources identified in the plans and measures to support their implementation, and a monitoring and evaluation system. evaluation
Emphasis on meeting economic and social needs on a sustainable basis	Shared concern with producers in need of economic and food satisfaction	The action plans drawn up do not give particular importance to productive investment, financial support for producers and the valorization of the productions	To find solutions that facilitate the reasoned increase of agro-sylvo-pastoral production and its economic value in the basin

Reference value in GDRN	Forces	Weaknesses	Areas for improvement
Technical methods			
Appropriate and enabling policy, legislative and regulatory framework	Organization of work in accordance with the provisions of the law	Low level of mastery, appropriation and application of the various texts described Non-permanent, insufficient means of exercise, not very adaptive, low level of technical and technological innovation, persistent sector spirit, laxity, relinquishing mobility, etc. Low level of appropriation and application of the texts governing decentralized management of natural resources, which is reflected in the absence of a willingness to strive for harmonization, coherence, and synergy of action in the formulation of structured and effective responses, sufficient and sustainable.	To improve the performance of technical services through training in initiative and innovation, and to improve the means and quality of their use

Planned and organized management of natural resources	Carry out within the framework of politically and institutionally pre-determined programs (centralized planning)	Lack of rigorous work plans, monitored and evaluated against defined and assigned objectives Insufficient synergy of action and low level of innovative initiatives Low level of initiative and innovation in the context of the new advisory support functions to be provided to local authorities	Improve the level of planning by objectives and implement a monitoring and evaluation system to measure and assess the efforts and results obtained in relation to the overall objectives sought Implement a strategy of resource mobilization and
Reference value in GDRN	**Forces**	**Weaknesses**	**Areas for improvement**
			results management items interventions
emphasis on meeting economic and social needs on sustainable basis	Is part of the global vision of supporting the economy through natural resource management, taking into account the economic priorities of the SCADD	On the economic level, priority is given almost exclusively to the agricultural and livestock sectors. Low level of adequacy between environmental resources and economic production. Low level of mastery of concepts related to the optimal use of natural resources, low level of promotion and support of producers in productive investments, low level of economic development and creation of added value of agro-sylvanical productions. pastoral	To train in the appropriation of sustainable production systems, the promotion of investments and economic activities alternative to the abusive exploitation of natural resources.

Methods of development projects and programs

Physical forest management methods			
Appropriate and enabling policy, legislative and regulatory framework	Translate sectoral policies defined at the national level into the field (Kou watershed)	Texts and laws on the subject are not sufficiently known and popularized among the populations, hence the occasional participation but not inscribed in the consciousness and behavior There is no predefined mechanism to ensure the continuity of the actions and the capitalization, valorization and dissemination of the results and the perpetuation of the achievements	Improve the conscious empowerment and participation of the main actors by making available to them all the texts translated and made accessible Automatically define a mechanism to ensure the continuity of actions, the capitalization, valorization and

Reference value in GDRN	**Forces**	**Weaknesses**	**Areas for improvement**
		Lack of an anticipated or functioning mechanism to identify, analyze, and mitigate outcome risks	dissemination of the results and the perpetuation of the acquired knowledge Always provide a mechanism for identifying, analyzing and addressing risks related to the results by linking the objectives pursued and the goal sought

Planned and organized management of natural resources	Elaboration, implementation and monitoring of development plans and integrated forest management based on the exploitation objectives of wood and non-wood forest products	Incomplete implementation of plans and lack of support or capacity to complete the required forest planning cycle; Lack of a proactive or functioning mechanism to identify, analyze and mitigate risks to outcomes	Guarantee the completion of the implementation of forest management plans through the mobilization of the necessary funds and the strengthening of the capacities of the GGF Always provide for a system for identifying, analyzing and dealing with risks related to results in relation to the objectives and the goal sought
Emphasis on meeting economic and social needs on a sustainable basis	The development and management plans for forest or pastoral areas include the satisfaction of economic and social needs by promoting the planned use of woody forest resources	Support fund for private initiative for the valorisation of wood and non-wood forest products Low level of promotion of alternative activities from legally harvested forest products Confusion of interests (foresters more interested in their own interests than in the integrity of forest resources)	Ensure adequate support in the time to implement forest management plans Facilitate initiatives of transformation and valorisation of forest products by the actors at the base. Always provide a device for identification, analysis and

Reference value in GDRN	Forces	Weaknesses	Areas for improvement
	and non-timber products as well as the promotion of income-generating activities from the collected products Consideration of sectoral economic interests (agroforestry, grazing controlled, etc.)	Lack of an anticipated or functioning mechanism to identify, analyze, and mitigate outcome risks	treatment of risks related to the results in relation to the objectives pursued and the goal sought
Methods of hydro-agricultural development			
Appropriate policy, legislative and regulatory framework	Translate sectoral policies defined at the national level into the field (Kou watershed)	Policy, strategic, legislative and regulatory instruments are not disseminated to producers There is no predefined system to ensure the continuity of actions and the capitalization, valorization and dissemination of results and the sustainability of achievements Lack of an anticipated or functioning mechanism to identify, analyze, and mitigate outcome risks	Disseminate water resources governance tools to producers Always provide for a system for identifying, analyzing and treating risks related to results, in line with the objectives and purpose Always provide for a system for identifying, analyzing and treating risks related to results, in line with the objectives and the goal sought

Planned and organized management of natural resources	Hydro-agricultural developments are planning instruments space management and	The survival of the facilities is not assured after the interventions are stopped	To make accessible and controllable the design and techniques of hydro-agricultural developments

Reference value in GDRN	Forces	Weaknesses	Areas for improvement
	hydro resources agricultural Use of a map resources	The cartographic supports of development are not always understood and mastered by the producers Absence of an advance device or lack thereof to identify, analyze and evaluate the Mitigate risks to results	Automatically predefine a device to ensure the continuity of actions, the capitalization, valorization and dissemination of the results as well as the perpetuation of the acquired knowledge Always provide a device identification, analysis and treatment of risks related to results in relation to the objectives and the goal sought
Emphasis on meeting economic and social needs on a sustainable basis	The hydro-agricultural developments are par excellence productivist and thus with strongly economic ends	The need for sustainability of the exploited resources is not always well perceived and transmitted to the producers in order to integrate it in their behaviors and actions The existence of a financial mechanism to accompany producers in the time is not assured	To take into account the parameters of sustainability in the systems of management and exploitation of the managed areas Automatically predefine a device to ensure the continuity of actions, the

		Absence of an advance device or lack thereof to identify, analyze and evaluate the Mitigate risks to results	capitalization, valorization and dissemination of the results as well as the perpetuation of the acquired knowledge Always provide a device identification, analysis and treatment of risks related to results in relation to the objectives and the goal sought

Reference value in GDRN	Forces	Weaknesses	Areas for improvement
Conservative water and soil management methods			
Appropriate policy, legislative and regulatory framework	Translate sectoral policies defined at the national level to the field (Kou watershed)	Policy, strategic, legislative and regulatory instruments are not disseminated to producers There is no predefined device for ensure the continuity of actions and the capitalization, valorization and dissemination of results and the sustainability of achievements Absence of an advance device or lack thereof to identify, analyze and evaluate the Mitigate risks to results	Disseminate water resources governance tools to producers Automatically predefine a device to ensure the continuity of actions, the capitalization, valorization and dissemination of the results as well as the perpetuation of the acquired knowledge Always provide a device identification, analysis and

35

			treatment of risks related to results in relation to the objectives and the purpose for which it is intended
Planned and organized management of natural resources	The facilities water and soil conservation (control of a land degradation, recovery and increase of cultivable areas) are tools for relevant to the planning of the use and management of the land resource	Conservative development actions of water and soil still do not cover the necessary living space of the populations The means and time for accompanying the actors the time to have full control of the techniques and technologies are generally insufficient Absence of an advance device or lack thereof to identify, analyze and evaluate the Mitigate risks to results	Cover all the spaces of production and use of resources through the development of water and soil conservatories Always provide a device identification, analysis and treatment of risks related to results in relation to the objectives pursued and the goal sought

Reference value in GDRN	Forces	Weaknesses	Areas for improvement
Emphasis on meeting economic and social needs on a sustainable basis	The conservation of water and soil (fight against land degradation, recovery and increase of cultivable areas) are excellent practices that support sustainable production controlled by the producers themselves. same	The means to accompany the planned exploitation of certain structures for ecological, social and economic purposes are not sufficiently available Lack of an anticipated or functioning mechanism to identify, analyze, and mitigate outcome risks	Always provide a system for identifying, analyzing and treating risks related to the results by linking them to the objectives pursued and the goal sought

From the above, it should be noted that the methods as varied as the stakeholders who use them have produced different recipes. There is something to be learned from the lessons of the past in order to better plan for the future in light of present realities and future challenges. However, no method, no matter how effective, can be successful if the institutional and organizational framework that promotes it remains dysfunctional or at least not permanent. The greatest shortfall in the search for a certain sustainability for the actions and for the biophysical and functional integrity of the natural resources of this basin lies not only in the non-permanence of the interventions but also in the near impossibility of linking in time and space the initiatives themselves and the results, experiences and achievements that could constitute the mainsprings of sustainable management. In natural resource management, however, any slackening destroys the fruits of efforts and extraverts the results and achievements. The functional links between results in time and space are difficult to establish, which is a source of their loss.

In short, the sustainability of natural resources and ecosystems depends as much on the relationship between humans themselves as on the relationship between humans and nature.

Table 7: Description of Report Types

Deficiencies noted	Nature of the reports		Description
	Actors/players	Actors/NR	
Sight steering of development policies and objectives	X		Questions the management of hierarchical relationships, definition and clarification of roles and responsibilities and mechanisms for achieving objectives
Inconsistency of interventions in space and time	X	X	Lack of synergy of action, complementarity, pooling of resources and experience, Failure to take into account the links between people, space and time
Ineffectiveness of consultation frameworks	X		Low level of management productivity: low rate of implementation of decisions and recommendations Problems of individual and collective responsibility Sustainable management of natural resources is not well perceived
Insufficient resource mobilization and promotion of productive investments	X	X	Lack of coherent policy and strategy for resource mobilization, investment and organization of production that enhances ecologically, socially and economically the natural resources of the basin

Institutional weaknesses in the management of results and achievements	X	X	Poor perception of the role and place of interventions in achieving development objectives, Under-appreciation of the importance of the results and achievements in relation to the sustainable management of natural resources
Persistence of poor human practices	X	X	Low level of ownership and application of rules and modes of access and of technologies and techniques for the use of natural resources Lack of human awareness of the ecological, social and economic importance of natural resources

The results of the analysis of the political, institutional and technical system in place show the need, indeed the urgency, to improve this system and its constituent elements by enhancing its strengths and correcting its weaknesses and shortcomings. Concretely, it is necessary, on the one hand, to work to make effective, efficient, permanent, productive, the relations between the men themselves around the responsibilities to be accomplished, the definition and the application of the policies, the legislative and regulatory texts, the respect of the obligations and on the other hand, to make sure that the links that the men maintain with the natural resources which must be managed on durable and constructive bases on all the ecological, economic, social and cultural plans, are healthy, reasoned links.

Thus, in order to bring together the determinants of successful management of the natural resources of the Kou watershed that ensures their biophysical and functional integrity, we propose a method of strategic governance of these resources. This method is based on an ecosystemic, ecological, economic and socio-cultural approach to natural resource management in the Kou watershed.

And to summarize everything, I propose this diagram on the sustainable management of natural resources. It is one of the bases of the strategic governance method developed following the various observations.

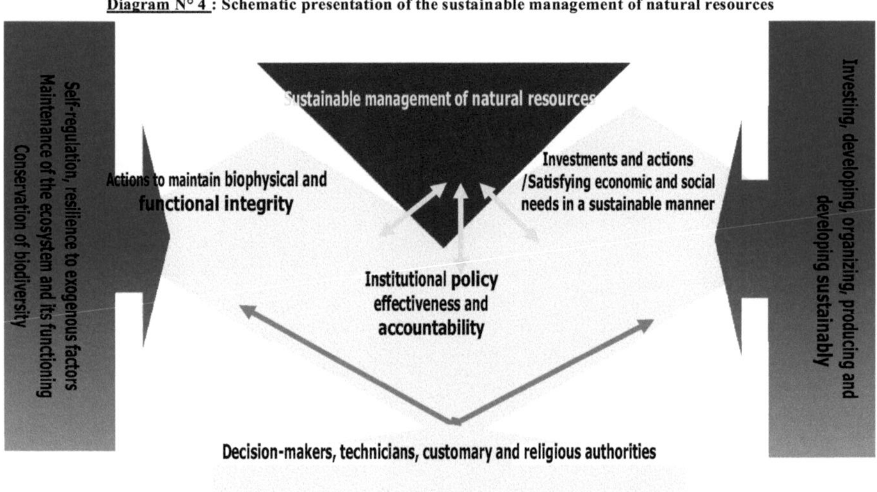

Diagram N° 4 : Schematic presentation of the sustainable management of natural resources

Table of contents

BIBLIOGRAPHY

<u>Sources</u>

Workshop on the constraints and difficulties of the rice-growing plain of Bama and proposed solutions for sustainable production, February 22 in Bama

BURKINA FASO, Ministry of Agriculture. Communication on the results recorded by the rural development sector in the fight against poverty; November 2009, p.2

The Convention on Biological Diversity (CBD) is a global treaty that was adopted at the Earth Summit in Rio de Janeiro in 1992. Its objective is to develop national strategies for the conservation and sustainable use of biological diversity. It is recognized as the key document for sustainable development.

African Convention for the Conservation of Biodiversity and Natural Resources. It was adopted on July 11, 2003 in Maputo, Mozambique: the conservation and management of natural resources must be treated as an integral part of national and/or local development plans. In the formulation of these plans, ecological, economic, social and cultural factors should be taken into account, adopted on 16

Document of the development and conservation management plan of the water and soil of the Farakoba sub-watershed;

Interview with SANON. A. Titianma, Bishop of Bobo, Interview conducted on November 06, 2012, Senufo Cultural Center;

Interview with COMPAORE Nestor, VREO Coordinator, February 17, 2013; Interview with

MidécorAkoly, AEDE, February 17, 2013;

JO AOF of August 3, 1935, p.611. Corrigendum JO AOF of September 14, 1935, p.723. Completed by the decree of April 12, 1954 (art. 23 bis), JO AOF of May 8, 1954, p.844. Modified by the decree n°55 582 of May 20, 1955 (JO AOF of June 11, 1955, p.1004-1006). General decree n°2195 S. E. of September 28, 1935 defining the southern limit of the Sahelian zone and regulating the exploitation of forests (JO AOF of October 12, 1935, p.797). The latter institutes a normative regime with repressive regulations. Indeed, more than half of the articles (43 out of a total of 84) relate to the repression of infractions (Title V, from Article 36 to Article 78).

Regional forum on the restoration and protection of the Kou river banks, 15-16 September 2011, Bobo-Dioulasso

Ouattara, F.B., Communication on *"Elaboration of local rules of natural resource management, theory and practice, the legal framework of the Samorogouan experience" on the* occasion of the capitalization workshop of the implementation of local rules of natural resource management in the department of Samorogouan, related to the management of the Centre d'Encadrement des Zones d'Intensification de l'Élevage Traditionnel (CEZIET), **Orodara, March 18-19, 2008;**

Development and management plans for the two classified forests of Dindéresso and Kou

Training reports on the Methodological Guide for the creation and management of conservation spaces by local authorities (provincial and communal levels);

Self-evaluation report on the activity programs of forest management groups, 2007

Self-evaluation reports on the implementation of annual village investment plans and farmers' organizations' activity programs.

Policies, strategies and plans

The "8000 villages, 8000 forests" and "one school, one grove" approaches,

Ministry of Economy and Development. DECREE N°2002- 604 /PRES/PM/MEDEV adopting the decentralized rural development policy letter;

Ministry of Agriculture. Rural Development Strategy (RDS) to 2015: It was developed in 2002 and adopted in 2003 as an instrument for implementing reforms in the agricultural sector including agriculture, livestock, forestry, etc. with the aim of increasing performance and productivity on a sustainable basis;

Ministry of the Environment. National forestry policy: It is based on the PNAF as a complement to the decentralized rural development policy letter (LPDRD, 2002) and as a contribution to the implementation of the rural development strategy document (SDR) adopted in December 2003;

Ministry of the Environment. Decree n°2007-460/PRES/PM/MECV/MFB of March 30, 2007 on the National Environmental Policy (PNE),

Ministry of Agriculture. Decree No. 2002-539/PRES/MAHRH/ of March 12, 2003 approving the action plan for integrated water resources management (PAGIRE) and Decree No. 2003-220/PRES/PM/MAHRH of May 6, 2003 adopting it.

Ministry of Agriculture. Decree No. 2007-610/PRES/PM/MAHRH, adopting the national policy on land tenure security in rural areas of 4 October 2007.

Ministry of Economy and Finance, General Directorate of Land Use, Local and Regional Development, Etude du schéma national d'aménagement du territoire du Burkina Faso,

Ministry of Economy and Finance. Ministry of the Environment and Sustainable Development, Burkina Faso Poverty-Environment Initiative Project (IPE/BURKINA), Guide to Integrating the Environment and Poverty-Environment Linkages into Sectoral Policymaking in Burkina Faso;

MINISTRY OF AGRICULTURE, HYDRAULICS AND RESOURCES
HALIEUTIQUES, Programme de valorisation des ressources en Eau de l'Ouest (VREO), Etude des périmètres de protection des sources de Nasso et des forage de l'ONEA, Rapport Final, 2009

MINISTRY OF AGRICULTURE, HYDRAULICS AND FISHERIES RESOURCES, General Secretariat, General Directorate for the Inventory of Hydraulic Resources, Local Water Committees, Design Guide Document creation and operation. 2004

The national prospective study "Burkina 2025": It has identified strategic orientations for the formulation of development policies and strategies;

the national policy on good governance (PNBG): It organizes and regulates the actions, acts and behaviors of actors. In its eighth objective, it recalls the need to "preserve and manage the environment and natural resources for sustainable development",

Plans, programs

National Forest Sector Program (PNSF), September 2011 version, page 8. The Action Plan for

Integrated Water Resources Management (PAGIRE),

The National Environmental Action Plan (NEAP), The Strategic Research Plan (1995),

The National Action Plan to Combat Desertification (1999),

The Operational Strategic Plan for Sustainable Growth of the Agricultural Sector (1999), The National

Framework for Climate Change Mitigation,

The Programme cadre de gestion durable des ressources naturelles et fauniques, a document resulting from the review of the Programme d'aménagement des forêts (PNAF), adopted in 1996;

The National Biodiversity Strategy and Action Plan (2000),

The Integrated Soil Fertility Management Action Plan (ISFMAP),

The Action Plan for the implementation of institutional and legal reforms for the decentralization of the forestry sector.

Agrarian and land reorganization (RAF) in Burkina Faso in 1984

The fight against bush fires, abusive wood cutting and animal roaming; The adoption of the National

Plan to Combat Desertification (PNLCD) in 1986.

Laws and regulations/agreements

Ministry of the Environment, Law N°002/97/ADP of 27 January 1997 on the Forestry Code;

Ministry of the Environment, La loi N°005/97/ADP du 30 janvier 1997 portant code de l'environnement ;

Ministry of the Environment, Law N°003-2011/AN of 05 April 2011, on the Forestry Code (Law N° 6/97/ADP of 31 January 1997 New);

Ministry of Animal Resources, Loi d'orientation n° 034/2002/AN du 14 novembre 2002 relative au pastoralisme au Burkina Faso

Ministry of Agriculture and Hydraulic Resources. Law No. 002-2001 on the orientation law for water management;

Ministry of Territorial Administration and Decentralization. Law n° 055/2004/AN, of December 21, 2004 on the General Code of Territorial Authorities (CGCT),

Ministry of Agriculture and Hydraulic Resources. Law n°014-94/ADP of May 1996 on agrarian and land reorganization in Burkina Faso

Burkina Faso, third legislature. Law No. 005-2006/AN of March 17, 2006 on the safety regime for biotechnology in Burkina Faso, published by the Ministry of the Environment and the Living Environment, April 2006

Ministry of the Environment and the Living Environment, Law n°003-2011/AN of 5 April 2011 on the Forestry Code in Burkina Faso

Ministry of Environment and Sustainable Development. Permanent Secretariat of the National Council for the Environment and Sustainable Development. Environment Plan for Sustainable Development, 2010

Ministry of Energy and Mines, Law n°033-2003/AN of May 8, 2003 on the mining code in Burkina Faso

Law n°022-2005/AN of May 24, 2005 on the general code of public hygiene in Burkina Faso

Law n°017-2006/AN of 18 May 2006 on the urban planning and construction code in Burkina Faso

Works

FERGUENE, Ameziane (Ed.), Gouvernance locale et développement territorial: le cas des pays du Sud, l'Harmattan, 2004

BROMLEY D., and CERNEA M., "Joint management of natural resources: some conceptual and operational errors, Washington, the Word Bank, 1989.

Cire Ba, B., 1930 and 1954, quoted by GUY LE MOAL, in "Bobo, nature et fonction des Masques",

KAMTO. M.,: Droit de l'environnement en Afrique; in Universités francophones, 1996.

DIALLO M., et al. inventory of CLE-KOU market gardeners, 2008:

CHATAUBRIAND, in "Extracts of quotations from his memoirs from beyond the grave", noted that "the forests precede us and the deserts follow us".

FRIEDMAN, J., Urban planning, 1969 and 1996;

GUY LE MOAL, BOBO, NATURE AND FUNCTIONS OF MASKS, OROSTOM, PARIS, 1980.

GWEY, M., Contribution des facultés de théologie catholique à la problématique du développement", Conference of October 17, 1983, in Revue théologique de Louvain, T. 17, 1984, p 191-202

JAUD, J.P., 2008, Documentary film "Nos enfants nous accuseront" (Our children will accuse us) on the subject of organic agriculture.

KAMTO, M., Droit de l'environnement en Afrique, edicef, Paris, p.56. Quoted by Sandrine Carole TAGNE KOMMEGNE, University of Limoges - Master 2 in international and comparative environmental law 2007, in "Gestion durable des ressources naturelles en Afrique Centrale: Cas des produits forestiers non ligneux au Cameroun et au Gabon".

KI-ZERBO, J., A quand l'Afrique? Interview with René Holenstein, Paris, Edition d'en bas, Ouagadougou, Sankofa and Gurli publishing.

LAZAREV, G ET ARAB, M, 2002, Développement local et communautés rurales. Approches et instruments pour une dynamique de concertation, Paris Karthala.

Lazarev, G; Local governance and sustainable management of natural resources: a common political response to change an unsustainable development model, Communication at the HCP. Maghreb Atlantic Forum, Skhrirat May 2009.

Leach Robert, Percy-Smith Janie; Local governance in Britain, Ed: Lavoisier, May 2001, 256p. Paperback State

LIPIETZ Alain, What is political ecology? La grande transformation du xxe siècle, Editions la découverte, 9 bis, rue, Abel-Hvelacque, PARIS XIIIe, 1999

PAMPHILE, S., November 2000, Acteurs et enjeux de la décentralisation et du développement local : Expérience d'une commune du Burkina Faso, ECDPM Discussion Paper n°21, 45p;

PAVAN, SUKHDEV, The Economics of Ecosystems and Biodiversity, Progress Report,

PECQUER, B. Le développement local. 2nd revised and expanded edition, Published on: 12 October 2000, Publisher: Syros, Collection: Alternatives économique. Bernard Pecqueur is a lecturer in economics at Pierre-Mendès-France University (Grenoble-II) and a Grenoble city councillor, where he chairs the "Economic Development and Employment" commission.

OECD, Natural Resources and Pro-Poor Growth: Economic and Policy Issues, 2009

SANOU D. B., Contribution de la coutume bobo-madarè à l'aménagement des forêts de Dindéresso et du kou dans la commune de Bobo-Dioulasso, p.38

SANON, G., Le monde comme dehors et dedans, Essai sur la philosophie madare, Thèse de 3e cycle, Strasbourg, 1980. Op.cit, P 128

SANOU, D.B, (under the direction of) Communautés villageoises de la province du Houet et projet BKF/007 .PAFDK. Un exemple de partenariat pour l'aménagement des forêts classées de Dinderesso et du Kou, CAD, 2004 p.29, 30, 31.

SANTAMARA, F., 2005: Article on local development. Frédéric Santamaria, is co-responsible for the lexicon project and a lecturer in urban planning and development at the University of Paris 7 Denis-Diderot.

SY, O., Reconstructing Africa: Towards a new governance based on local dynamics. Edition: Charles Léopold Mayer, 2009, p.146.

SAURET.E, 2007, Contribution to the understanding of the hydrogeological functioning of the aquifer system in the Kou basin. Works of DEA 2007.

Varone F., Reynard E., Kissling-Näf I. and Mauch C., 2002, "Institutional Resources Regimes. The Case of Water in Switzerland", Integrated Assessment. An *International Journal*, vol. 3, n° 1, pp. 78-94.

Memories

MIDEKOR, A. A., *Cartographie des utilisations du sol et des ressources en eau et analyse des perspectives associées à l'élevage dans le bassin du Kou,* DEA research paper, July 2009.

YE Ouanki Lucien, *Coopération décentralisée franco-burkinabé pour le développement durable des collectivités territoriales au Burkina Faso : Rôle des acteurs locaux à travers le jumelage Fosses-Kampti.* Master II thesis in Project Management, 2010.

Reports

Luxembourg Agency for Development Cooperation. Annual activity report, 2009

World Bank, World Human Development Report: Sustainable Development in a Dynamic World, Improving Institutions, Growth and Quality of Life, 2003

Martin Sommer, *preface to the Center for Development and Environment (CDE) report "Development and Environment No. 19",* 2004.

Ministry of Economy and Development, SNAT Study Report, 2008.

Hartwick, John M. [1977] *"Intergenerational Equity and the Investment of Rents from Exhaustible Resources" American Economic Review, 67, December, pp. 972-74.*

United Nations Development Programme (UNDP): Sustainable Human Development Report, 2011.

UNEP, State of the Environment Report (1972-1992)

SUKHDEV, P., TEEB (The Economics of Ecosystems and Biodiversity) Final Report. An economic valuation of biodiversity. Published on Monday, November 22, 2010 at 12:00 "

SOMMER, M. (2004), preface to the Center for Development and Environment (CDE) Report "Development and Environment No. 19".

UNEP, Conference of African Ministers of the Environment, Dakar, 2006.

Republic of Senegal, Ministry of the Environment and Nature Protection, Directorate of Water, Forests, Hunting and Soil Conservation, Support Project for the National Rural Forestry Program of Senegal, Origins, Political Context, Objectives and Approach, Dakar, March 1999

Electronic references

An article from Wikipedia, modified on June 7, 2011

Iratxe Calvo-Menda, " Analyse du régime institutionnel des ressources en eau : le cas du bassin versant de l'Andomarois ", *Développement durable et territoires*, [en ligne], Dossier 6 : les territoires de l'eau, uploaded on 10 February 2006, accessed on 06 March 2011. URL : http:// développementdurable.revue.org/1734.

Watersheds : Definition and explanations on www.technosience.net/? tab=glossary&definition=1132

I want morebooks!

Buy your books fast and straightforward online - at one of world's fastest growing online book stores! Environmentally sound due to Print-on-Demand technologies.

Buy your books online at
www.morebooks.shop

Kaufen Sie Ihre Bücher schnell und unkompliziert online – auf einer der am schnellsten wachsenden Buchhandelsplattformen weltweit! Dank Print-On-Demand umwelt- und ressourcenschonend produziert.

Bücher schneller online kaufen
www.morebooks.shop